# Public Lands
# Belong to You!

The Bureau of Land Management (BLM) is a federal government agency that takes care of more than 245 million acres of land. Most of these lands are in the western part of the United States. These lands are America's public lands, and they belong to all Americans. These public lands are almost equal in area to all the land in the states of Texas and California put together.

The BLM manages public lands for many uses. The lands supply natural resources, such as coal, oil, natural gas, and other minerals. The lands provide habitats for plants and animals. People enjoy the big open spaces on the lands. The lands also contain evidence of our country's past, ranging from fossils to Indian artifacts to ghost towns.

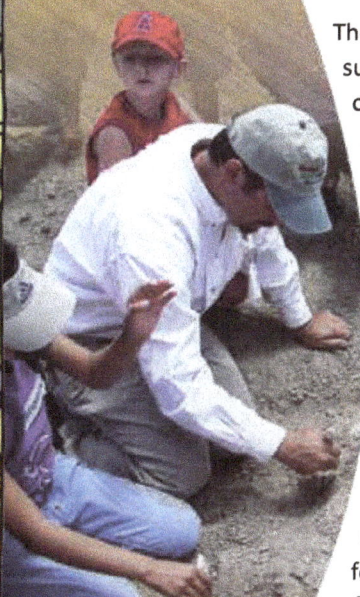

# Junior Explorers

BLM's Junior Explorer program helps introduce young explorers like you to the lands and resources that the BLM manages. This "Geology and Fossils Activity Book" focuses on nonrenewable natural resources. These are Earth features that formed over long periods of time and that cannot be replaced as humans remove and make use of them.

You can work through the activities on your own or invite a parent or an adult you know to join you. When you complete the activities, check them against the Answer Key in the back of the booklet. Then say the Junior Explorer pledge on page 21, sign the certificate, and you're on your way to exploring and protecting America's public lands. **Have fun!**

# Did You Know?

A geologist is a scientist who studies the Earth's structure and history. Much of a geologist's work focuses on the study of rocks and minerals. A paleontologist is a scientist who studies past life by looking at plant and animal fossils.

In this booklet you'll learn about rocks, minerals, and fossils. The booklet is full of fun activities, diagrams, and photos of BLM locations that illustrate many of the features you will learn about. You will also explore why geologic resources on your public lands are important and why we need to protect and use them wisely. The glossary on page 24 will help you understand these ideas better by explaining what certain words mean.

Lunar Crater National Natural Landmark, NEVADA

Lance Creek Fossil Area National Natural Landmark, WYOMING

2

# Fun Facts

The BLM manages many sites across the country where you and your family can explore and observe nature's geological wonders.

**ALASKA:** Hikers on *Alaska's Pinnell Mountain National Recreation Trail* will walk across some of the state's oldest rocks. Schist is the main rock type along the trail. Schist (a type of metamorphic rock) forms the prominent tors (rocky peaks) jutting from narrow ridge tops, as shown in the photograph to the left. This elaborately folded and deformed rock is between 700 million and 2 billion years old.

**UTAH:** The 1.9-million-acre *Grand Staircase–Escalante National Monument* (right photo) is so remote that it was one of the last places in the continental United States to be mapped. One of its most interesting features is a thousand-mile maze of interconnected canyons.

**ARIZONA/UTAH:** At the *Paria Canyon-Vermilion Cliffs Wilderness Area* (left photo), visitors can explore and walk through Buckskin Gulch, one of the longest and deepest slot canyons in the world. Slot canyons are sculpted in sandstone or limestone by moving water. Some slot canyons are only 3 feet across, but they are hundreds of feet deep.

**IDAHO:** *Craters of the Moon National Monument* (right photo) is co-managed by the BLM and the National Park Service. The monument is a vast open area that was named for its resemblance to the surface of the Moon. The site features lava flows and volcanic caves from volcanic eruptions that occurred within the last ten thousand years. Astronauts visited the site in 1969 to study volcanic geography before flying the *Apollo* Moon missions.

# More Fun Facts

**NEVADA:** Nevada's *Lunar Crater Volcanic Field* is a National Natural Landmark. The field looks so much like the surface of the Moon that NASA used it to train the *Apollo* astronauts. Lunar Crater was formed by a volcanic eruption and is the most distinctive feature of the field. The crater is approximately 426 feet deep and 3,444 feet wide (that's more than 10 football fields laid end to end)!

Nevada is also home to *Sand Mountain*. Sand Mountain is 600 feet high and is one of only a few "singing" sand dunes in the world. Winds moving across the dune rub the unusually shaped sand grains against one another. This action causes an eerie, moaning sound that can be heard if one is close enough.

# Geology on Your Public Lands

Many awesome shapes and forms of rocks and fossils are found on your nation's public lands. Some of these places are marked on the map on this page. Want to learn more about each of these unique sites? Visit **www.blm.gov.**

**Directions:** Write the name of the state on each line where a BLM site is found.

### POMPEYS PILLAR NATIONAL MONUMENT

This massive sandstone outcrop is the only major landform in the area. It has the only known physical evidence of the Lewis and Clark Expedition. Lewis and Clark were the first American explorers to travel across the United States and reach the Pacific Ocean. On July 25, 1806, during the trip back from the Pacific coast, Clark carved his signature and the date into the rock.

### ALABAMA HILLS

The rounded, weathered shape of the Alabamas appears very different from the sharp-looking ridges of the nearby Sierra Nevada Mountains. The rock formations have been the setting for many movies and television commercials.

### RED GULCH DINOSAUR TRACKSITE

More than 1,000 dinosaur tracks and 125 dinosaur trackways were discovered at this site. Visitors can closely examine the 167-million-year-old fossils.

1. _____

2. _____

3. _____

4. _____

5. _____

6. _____

7. _____

### GARDEN PARK FOSSIL AREA

Many famous dinosaur fossils, including fossils of *Allosaurus, Ceratosaurus, Diplodocus,* and *Stegosaurus,* have been found here.

### RED ROCK CANYON

The red- and cream-colored sandstone cliffs have been used as a scenic backdrop for more than 100 movies since the 1920s. This site is also home to a recently discovered *Grallator* dinosaur trackway.

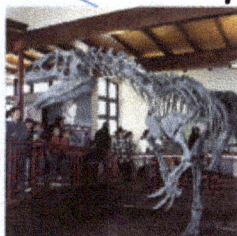

### CLEVELAND-LLOYD DINOSAUR QUARRY

This quarry contains the largest collection of Jurassic dinosaur bones in the world. An *Allosaurus* skeleton is on display in the visitor center.

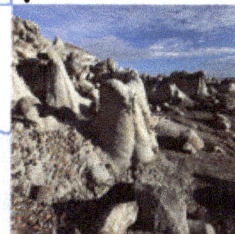

### BISTI/DE-NA-ZIN WILDERNESS

This area was home to large Cretaceous dinosaurs and reptiles 65–80 million years ago. The fossils provide a record of changes in plant and animal life at the end of the Age of Dinosaurs.

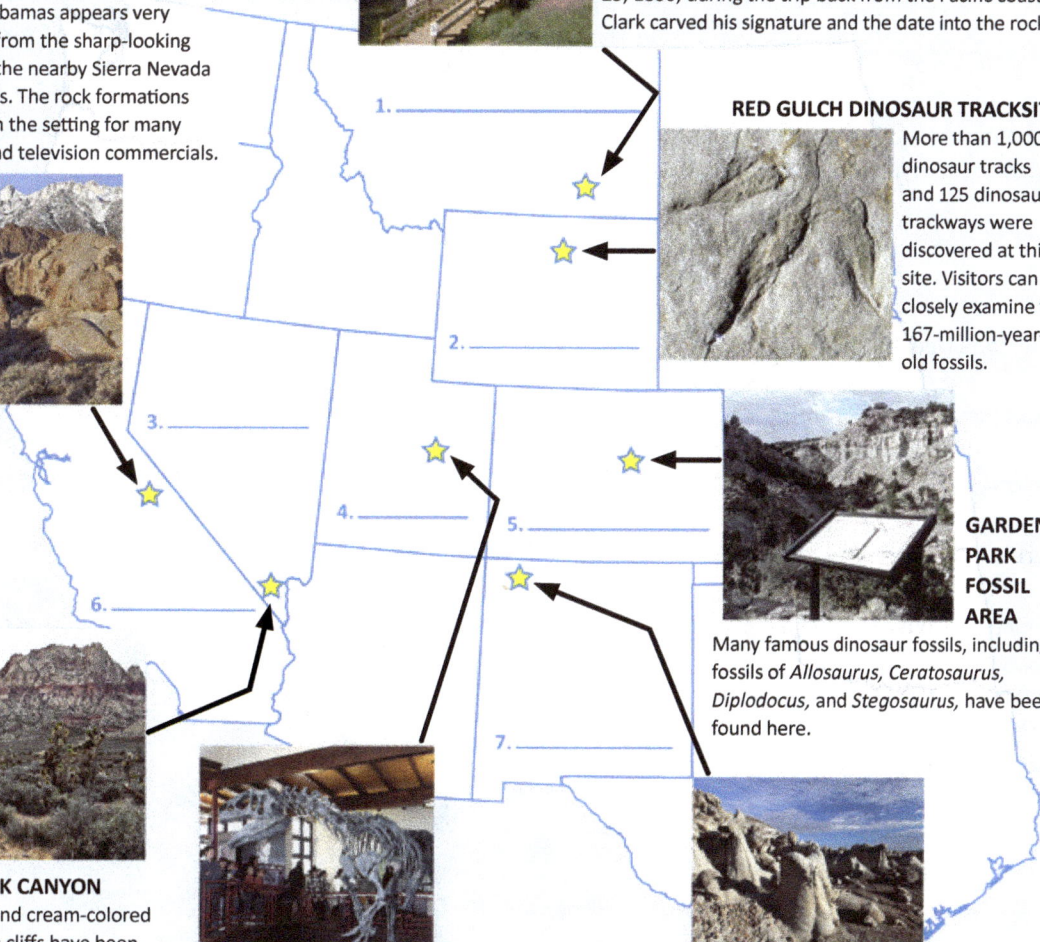

# Let's Rock!

What is a rock? A rock is composed of one or more minerals. Minerals are nonliving materials that are found in nature and that are chemically the same all the way through. Copper, diamond, gold, lead, pyrite, mica, and quartz are all minerals.

There are three kinds of rocks. Each kind is named for the way that it was formed.

**IGNEOUS ROCK** is formed when magma—molten (melted) rock deep within the Earth—solidifies or when it erupts through volcanoes on the Earth's surface as lava and then solidifies.

**SEDIMENTARY ROCK** is formed when layers of sand, clay, silt, or gravel settle and harden over time.

**METAMORPHIC ROCK** is formed when heat and pressure cause existing rocks to change slowly over time.

**Directions:** Match each rock type with the picture that represents how it was formed. There will be more than one rock type for each formation picture. The first one has been done for you.

Sandstone (sedimentary)

Basalt (igneous)

Limestone (sedimentary)

Marble (metamorphic)

Pumice (igneous)

Slate (metamorphic)

Coal (sedimentary)

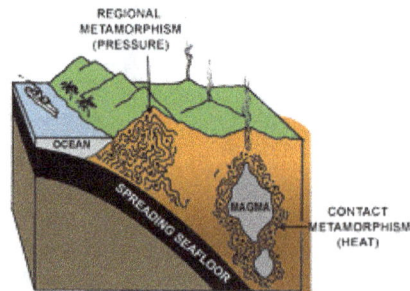

REGIONAL METAMORPHISM (PRESSURE)

OCEAN

SPREADING SEAFLOOR

MAGMA

CONTACT METAMORPHISM (HEAT)

Do you recognize the names of any of these rocks?
Where in your house or neighborhood have you seen or heard of examples of these rocks?

_____

_____

_____

# Rockin' Around the Rock Cycle (Part 1)

Geologists use a model called the rock cycle to help explain how rocks are formed and change over time. Weathering, erosion, heat, and pressure cause changes in rocks and minerals.

Weathering is caused by wind, ice, or water that breaks rocks into smaller and smaller pieces. Erosion occurs when moving wind, water, or ice wears away weathered rocks. Pressure is produced when layers and layers of rock push down on the Earth's crust. Rocks are heated when they are close to a heat source such as lava or magma.

**Directions:** Explore the rock cycle model, and then fill in the blanks in the descriptions below.

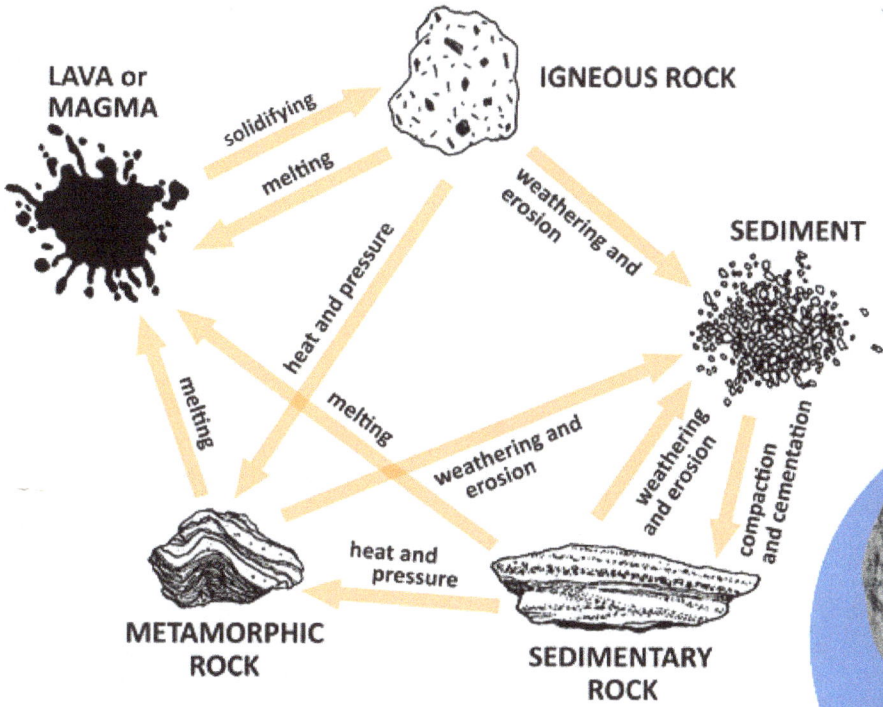

LAVA or MAGMA — solidifying / melting — IGNEOUS ROCK — weathering and erosion — SEDIMENT — heat and pressure — melting — melting — weathering and erosion — weathering and erosion — compaction and cementation — METAMORPHIC ROCK — heat and pressure — SEDIMENTARY ROCK

1. _____ rock is formed when _____ or
_____ rock is heated and under pressure.

2. Weathering and erosion create _____, which compacts and cements
to form _____ rock.

3. When _____ or _____ or _____ rock
melts, _____ or _____ is created.

4. When _____ or _____ solidifies, _____ rock is created.

# Rockin' Around the Rock Cycle (Part 2)

Think about how events in the environment could affect the formation of rocks in the rock cycle. For example, the movement of a heavy glacier could cause weathering and breakdown of igneous rocks into small particles. Then, over time, these particles might compact and harden into a sedimentary rock. Or, an earthquake could create pressure that changes igneous rock into metamorphic rock over time.

**Directions:** Select some of the environmental events from the illustrations below, and write them in the table under the effects that they can cause. Some events will fit under more than one heading.

| Weathering or Erosion | Heat and Pressure | Melting or Solidifying |
|---|---|---|
| *Wind* | | |
| | | |
| | | |
| | | |
| | | |
| | | |
| | | |

## Environmental Events

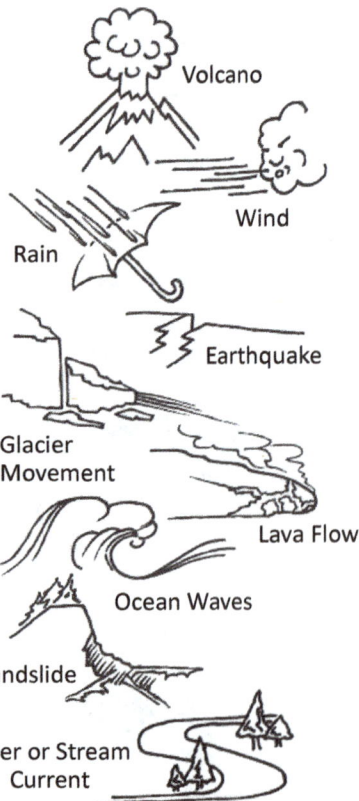

Volcano

Wind

Rain

Earthquake

Glacier Movement

Lava Flow

Ocean Waves

Landslide

River or Stream Current

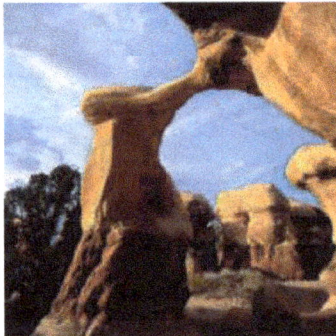

Metate Arch, Grand Staircase–Escalante National Monument, **UTAH**

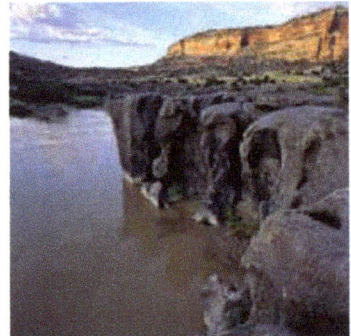

Black Ridge Canyons Wilderness, **COLORADO**

Kasha-Katuwe Tent Rocks National Monument, **NEW MEXICO**

"The Wave," Coyote Buttes, Paria Canyon Wilderness Area/Vermilion Cliffs National Monument, **ARIZONA**

# Rockin' Shapes and Forms

The weathering and erosion processes of the rock cycle can cause many different shapes and sizes of rock formations to develop. Softer rocks are more easily worn away by the effects of weathering. Water freezing and thawing can cause cracks or breaks in rocks. Wind can carry sand or pebbles. The sand and pebbles carried by the wind can hit rock surfaces and wear the rock into shapes such as arches or pillars. Running water can erode rocks, eventually carving canyons or creating underground caves. Weathering can also be caused by plant growth, which can crack rocks and break them down.

**Directions:** Complete the crossword puzzle below. Then unscramble the letters in the marked boxes to answer the question at the end of the activity.

## ACROSS

2. A _____ is a combination of minerals.

3. _____ carries sand and pebbles through the air.

4. _____ is a small particle carried through the air that wears away rock surfaces.

5. To _____ is to melt ice.

6. Water _____ can cause cracks and breaks in rocks.

8. An _____ is a curved structure created by wind erosion.

9. A _____ is a tall, vertical structure created by wind erosion.

10. A _____ is carved by running water.

## DOWN

1. _____ occurs when moving wind, water, or ice wears away weathered rocks.

3. _____ is a liquid that carves and creates rock features.

6. Weathering and erosion create many new rock _____.

7. When water freezes it turns into _____.

10. Running water might help create an underground _____.

What is the process of breaking down rocks by wind, water, and ice called?

JUNIOR EXPLORER FOSSILS GEOLOGY

FOSSILS

Activity Book

GEOLOGY

Activity Book

JUNIOR EXPLORER

# It's Classified

One way geologists study and organize rocks is by classification. They examine and describe rock textures, colors, hardness, and other characteristics.

**Directions:** Take some time to explore outside with an adult you know, and look closely at some rocks along your journey. Try to find rocks with different colors, textures, and shapes. Carefully study all of your rock samples, and describe them using the table below.

**Hint:** When describing your rocks, consider using words like **ROUGH, SMOOTH, BUMPY, SPOTTED, SHARP, FLAT, ROUND, SHINY, DULL,** or **MARBLED.** Be creative in thinking about the appearance and texture of your rocks.

|         | Color | Shape | Texture | Markings/Patterns |
|---------|-------|-------|---------|-------------------|
| Rock #1 |       |       |         |                   |
| Rock #2 |       |       |         |                   |
| Rock #3 |       |       |         |                   |
| Rock #4 |       |       |         |                   |
| Rock #5 |       |       |         |                   |

**Exploration Tip:**

It is important to remember to be a good steward of natural resources when you are exploring your public lands.

Ask a park ranger or another responsible adult if it is okay to touch or disturb rocks or fossils. Some resources should not be disturbed, and it is illegal to pick up certain kinds of fossils. Scientists use these resources to study the environment and the history of the Earth. Leaving the resources undisturbed also allows other visitors to enjoy viewing them later.

Instead of taking a resource home with you, take a photo, draw a picture, or write a poem or story about what you experienced on your journey outside.

# Lay it on Me: Geologic Layers

Geologists discovered that as sediment built up over long periods of time, layers of earth and rock formed. Geologists can "read" these layers (similar to "reading" the rings of a tree) and learn about the age of the rock formations. Sediment layers are built up from the bottom, so the highest layers of rock and soil are usually the youngest, and the lowest layers are usually the oldest. Geologists call the study of rock layers stratigraphy [struh-**tig**-ruh-fee].

Generally, rock layers form in flat rows. But when environmental events such as weathering, erosion, volcanic eruptions, or earthquakes occur, rock layers can become wavy or slanted, or they can be worn away to form caves or canyons. Younger layers of rock may also cut through or into one or more layers of rock.

These two diagrams demonstrate the oldest and youngest rocks in each rock formation.

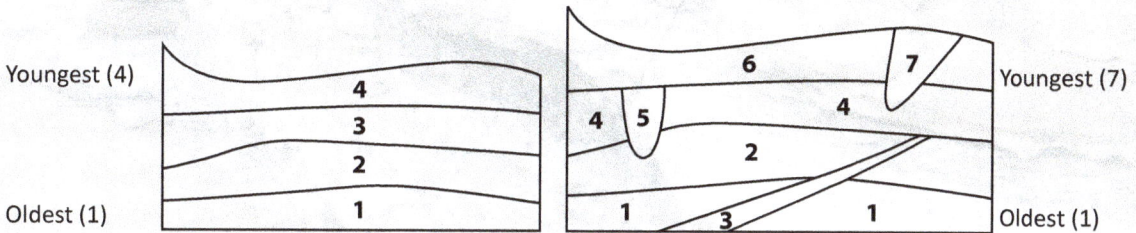

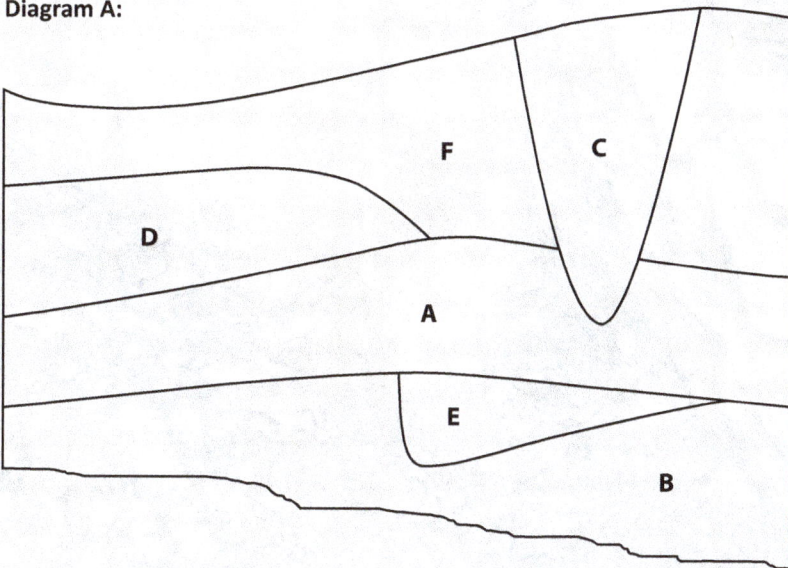

Youngest (4) — 4 / 3 / 2 / 1 — Oldest (1)

Youngest (7) — 6 / 7 / 4 / 5 / 4 / 2 / 1 / 3 / 1 — Oldest (1)

**Directions:** Study the diagrams above, and based on what you've learned about geologic layers, order the layers in the next two diagrams from oldest to youngest.

**Diagram A:**

Youngest — _____
_____
_____
_____
_____
Oldest — _____

Labels: F, C, D, A, E, B

# Lay it on Me: Geologic Layers (continued)

**Diagram B:**

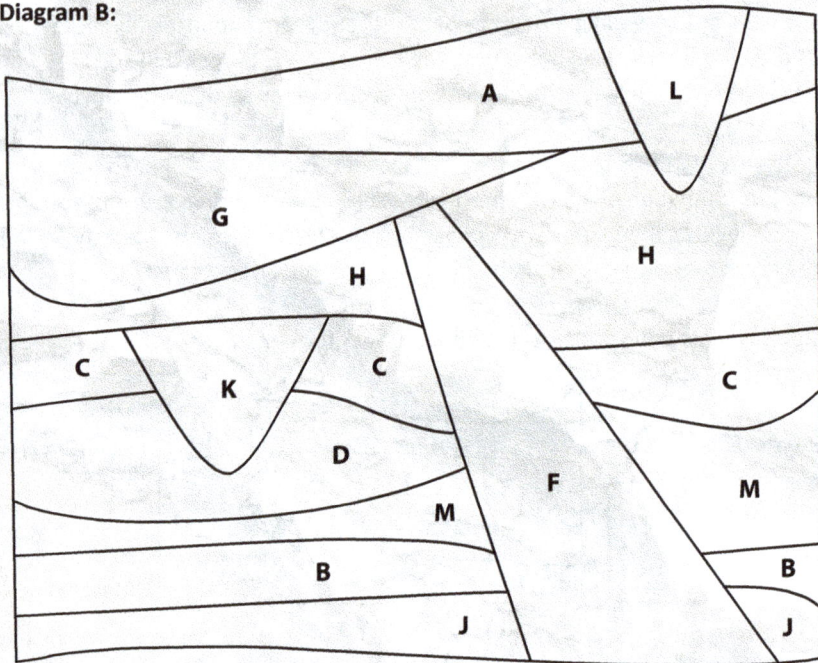

Youngest   ————

————

————

————

————

————

————

————

Oldest   ————

**Take It Outside!**

Go for a short hike with an adult you know, and look for evidence of geologic layers in hillsides, cliffs, rock formations, highway cuts, or other landscape features. Can you see evidence of erosion or weathering? Draw some pictures of your findings on the pages that follow.

# Lay it on Me: Geologic Layers (continued)

**Directions:** Draw some pictures of what you saw on your hike.

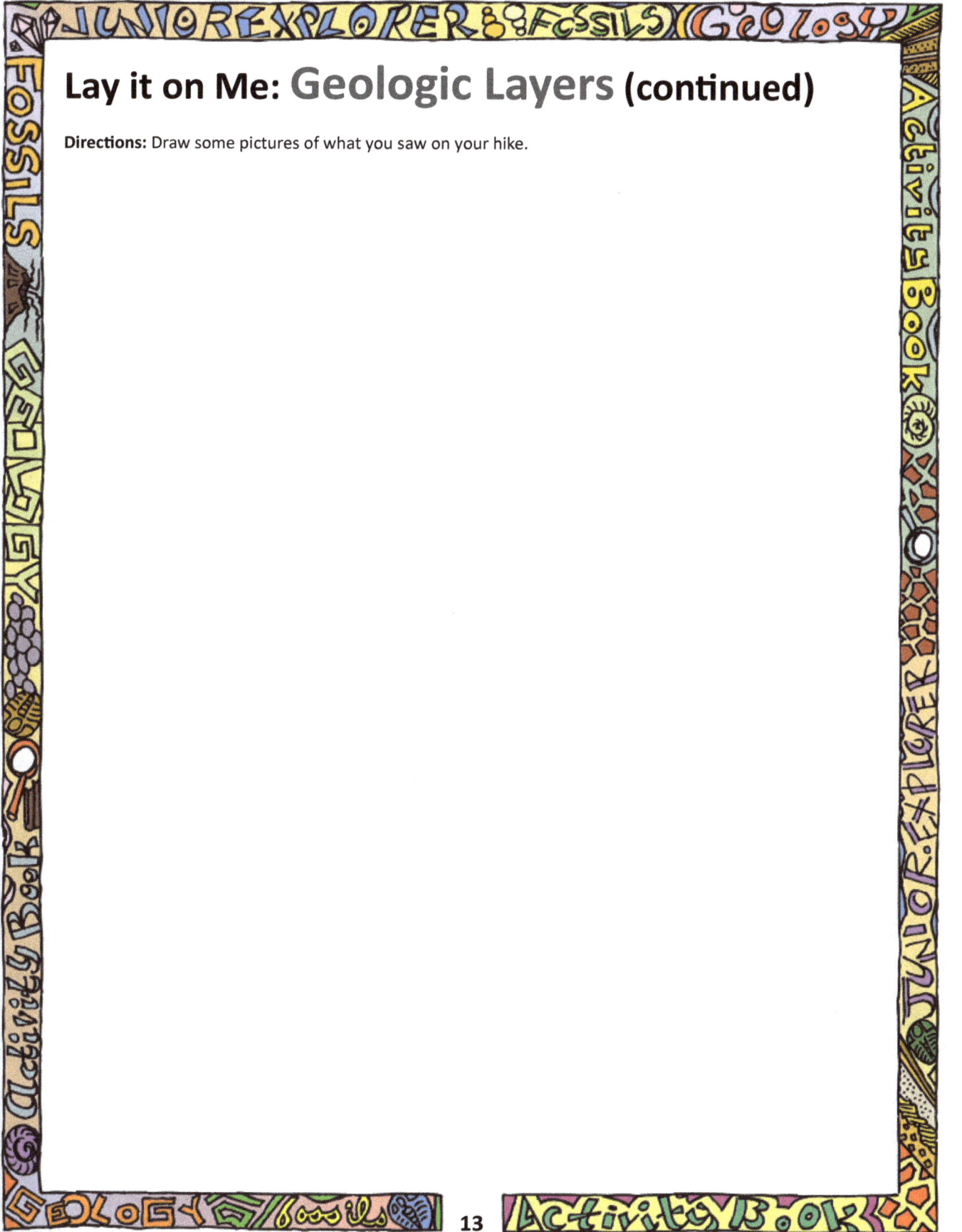

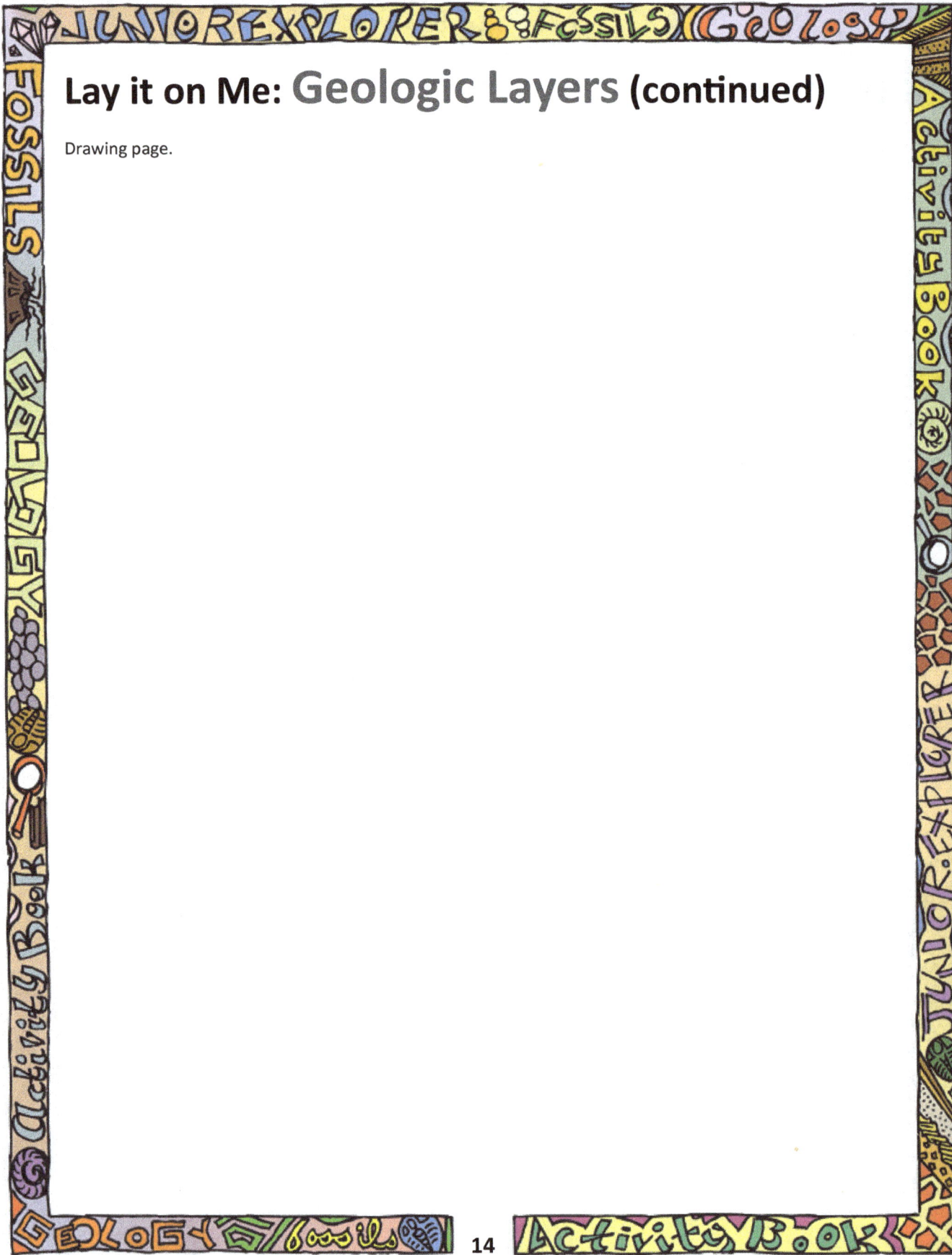

# Lay it on Me: Geologic Layers (continued)

Drawing page.

# History in the Rocks!

Fossils are the traces or evidence of once living things within layers of rock. They might be plants, fish, shells, animals, insects, or even footprints. Some fossils form when minerals replace organic material over time. Fossils offer one of the best ways for scientists to study the history of the Earth and nature.

The locations of fossils within the layers of rock help us understand the age and development of living things long ago. Fossils also help us understand how the Earth's environment has changed over long periods of time.

**Directions:** Cut out the fossils represented in the drawings below, and paste them on the next page with the habitats where they might once have lived.

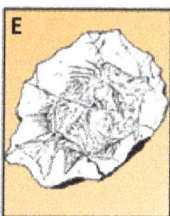

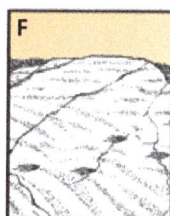

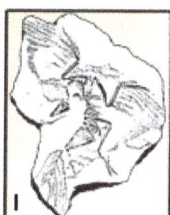

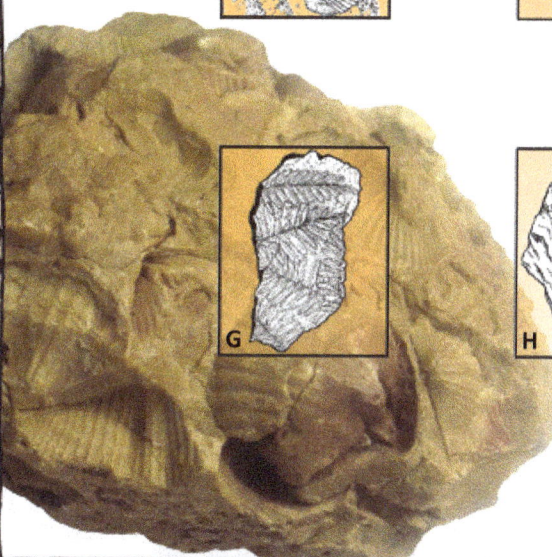

A

B

C

D

E

F

G

H

I

# Cut and
## paste

# History in the Rocks! (continued)

Forest

Ocean

Tidal Flat

# Being a Good Steward

An important role you can play in protecting your public lands is to be a good steward.

Stewardship means the careful and responsible treatment of something entrusted to your care. We are all responsible to care for the nation's public lands and resources so that they are here for future generations to visit, study, and enjoy.

Some ways that you can participate in caring for your public lands are by keeping them litter-free and by being careful not to cause damage to special resources. The rocks and fossils found on your public lands are some of those special resources.

**Directions:** Look at the pictures below. Circle the actions that help to protect special resources. Draw an X through the pictures that show actions that would harm special resources.

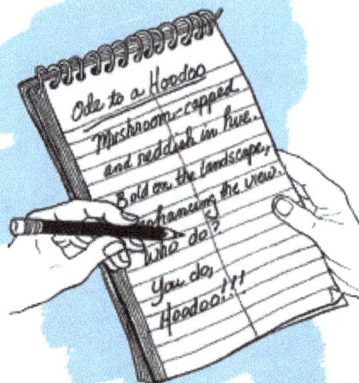

Can you think of other ways to help protect your public lands?
Share your ideas with someone you know!

# Career Profile
## Jim Goodbar

**Senior Cave and Karst Resource Specialist, New Mexico**

**Education**—Jim went to college and graduate school and studied park and recreation management as well as cave and karst geology. (Karst is a landscape that includes such features as caves, sinkholes, and underground rivers.)

**BLM Career Highlights**—Much of Jim's career with the BLM has been with BLM's Cave and Karst Management Program. Jim helped create instruction manuals about cave management, and he worked on four national laws to protect and study caves.

*Discovery!*—In 1991 Jim discovered the skeleton of a giant short-faced bear in a New Mexico cave. He found the bones piled at the base of a 180-foot drop-off. The huge bear, a fierce predator that's now extinct, lived between 10,000 and 30,000 years ago. Jim's discovery was the first time that a complete skeleton of this species had been found in the Southwest.

*Creativity*—In the summer of 1997 Jim designed, built, and staffed a simulated cave for the BLM's exhibit at the National Boy Scout Jamboree in Fort A.P. Hill, Virginia. The cave featured a "bat nursery," cave formations, an underground stream, cool breezes, and bilingual information plaques along three winding corridors. Jim also designed an inflatable cave for Boy Scouts to explore at the 2010 Boy Scout Jamboree.

*Rescue*—Jim has served as a New Mexico State Police search and rescue field coordinator and as an emergency medical technician. In November 1991 Jim received the BLM's Caren Padilla Memorial Safety Award for serving as the operations chief during an intense 4-day rescue in New Mexico's Lechuguilla Cave. An expert caver had broken her leg when she was more than a mile into the cave and a thousand feet deep. She was safely rescued by Jim and a 70-person team.

**At Home**—When he's not working, Jim is never far from his love of the underground world. Jim enjoys caving with his wife and son in the Guadalupe Mountains near their home in Carlsbad, New Mexico. As Jim says, "You can't know, unless you go!"

American Short-Faced Bear Skeleton, NABC/www.bear.org

Fort Stanton-Snowy River Cave National Conservation Area, **NEW MEXICO**

# Bureau of Land Management
## Junior Explorer

As a Bureau of Land Management Junior Explorer, I promise to:

- do all I can to help preserve and protect the natural and cultural resources on our public lands,

- be aware of how my actions can affect other living things and the evidence of our past,

- keep learning about the importance of nature and our heritage, and

- share what I have learned with others!

_____          _____
         Date                                    Explorer Signature

# Cut out and
## save certificate

# Answer Key

**Geology on Your Public Lands (page 5):**
1. Montana; 2. Wyoming; 3. Nevada; 4. Utah;
5. Colorado; 6. California; 7. New Mexico

**Let's Rock (page 6):**

| | |
|---|---|
| Volcano: | basalt and pumice |
| Sedimentary Layers: | sandstone, limestone, and coal |
| Heat and Pressure: | marble and slate |
| Sample Rock Uses: | marble floors, coal-burning furnace, slate patio, pumice bath accessories, sandstone fireplace, limestone roadbed |

**Rockin' Around the Rock Cycle (pages 7–8):**
1. metamorphic; igneous or sedimentary
2. sediment; sedimentary
3. igneous or metamorphic or sedimentary; lava or magma
4. lava or magma; igneous

| Weathering or Erosion | Heat and Pressure | Melting or Solidifying |
|---|---|---|
| Wind | Earthquake | Volcano |
| Glacier Movement | Glacier Movement | Lava Flow |
| Landslide | Volcano | |
| Ocean Waves | | |
| Rain | | |
| River or Stream Current | | |

**Rockin' Shapes and Forms (page 9):**
Unscrambled answer: weathering

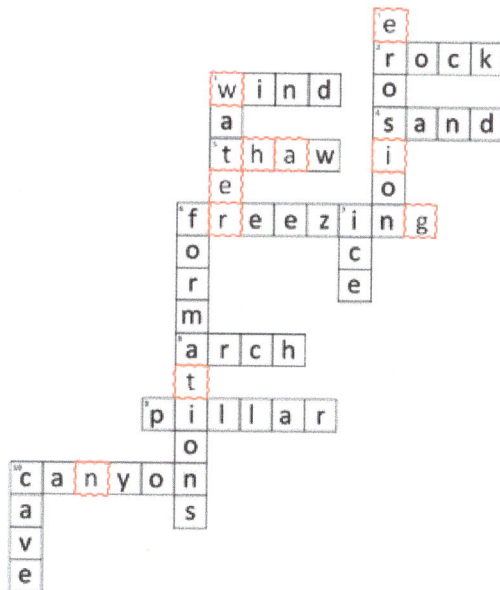

**Lay It on Me: Geologic Layers (pages 11–12):**
Diagram A: B (oldest), E, A, D, F, C (youngest)
Diagram B: J (oldest), B, M, D, C, K, H, F, G, A, L, (youngest)

**History in the Rocks (pages 15–17):**
Forest: A, G, I; Ocean: B, C, E; Tidal Flat: D, F, H

**Being a Good Steward (page 18):**
*Circle:* taking a picture of a scenic landscape, learning from a park ranger about public lands, painting a picture of a scenic landscape, writing a poem about your visit to public lands
*X:* taking home a fossil, writing or drawing on natural resources

23

# Glossary

**erosion:** the removal of material by water, wind, or ice

**fossils:** the traces or mineralized remains of plants or animals or other living things

**fossil trackway:** the fossilized footprints on an ancient route of travel

**geocaching:** an outdoor adventure game that uses a Global Positioning System (GPS) to hide and seek hidden containers, or geocaches, which hold trinkets for trading or log books

**geologist:** a scientist who studies the history and structure of the Earth

**highway cut:** part of a mountain or hill that is cut out to make way for a highway

**igneous rock:** rock formed from magma or lava that has cooled and hardened

**lava:** molten rock that flows from a volcano on the surface of the Earth

**magma:** molten rock below the surface of the Earth

**metamorphic rock:** igneous or sedimentary rock that has been changed through heat and pressure

**mineral:** a solid that is made up of specific chemicals and generally forms crystals

**molten:** changed into liquid by heat

**NASA:** National Aeronautics and Space Administration

**organic material:** material that comes from a living thing

**rock cycle:** a model showing how rocks are formed, changed, destroyed, and reformed

**sedimentary rock:** layers of sand, clay, minerals, or gravel that harden into rock; often contains fossils

**silt:** loose particles of rocks or minerals finer than sand but coarser than clay

**slot canyon:** a very narrow canyon that is much deeper than it is wide

**stewardship:** the careful and responsible treatment of something in your care

**stratigraphy:** the study of sedimentary rock layers

**weathering:** a process that breaks rocks down by wind, water, or ice

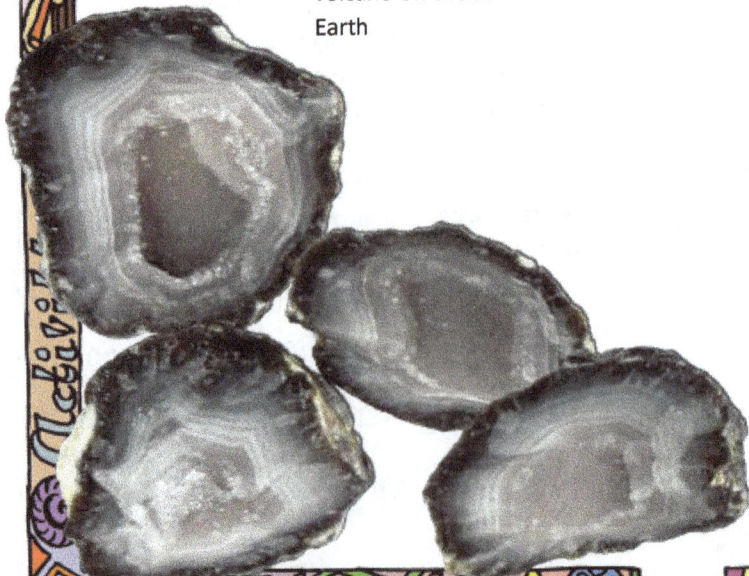

JUNIOR EXPLORER FOSSILS GEOLOGY

Activity Book

JUNIOR EXPLORER

GEOLOGY fossils Activity Book

Activity Book

GEOLOGY

FOSSILS

www.ingramcontent.com/pod-product-compliance
Lightning Source LLC
Chambersburg PA
CBHW080537030426
42337CB00023B/4777